E R... Y of GREENW...

WITHDRAWN FROM UNIVERSITY OF GREENWICH LIBRARY

UNIVERSITY OF GREENWICH
0 4 OCT 1999
LIBRARY FA

Study ban

UNIVERSITY OF GREENWICH LIBRARY
641.
588
2
HIL

D0198447

# MICROWAVE OVENS

Text developed by Allison Hill

*in collaboration with*

The ILSI Europe Microwave Oven Task Force

## ILSI
### International Life Sciences INSTITUTE

# ILSI Europe

© 1998 International Life Sciences Institute

*All rights reserved.* No part of this publication may be reproduced, stored in a retrieval system, or transmitted, in any form or by any means, electronic, mechanical, photocopying, recording, or otherwise, without the prior written permission of the copyright holder. The International Life Sciences Institute (ILSI) does not claim copyright on U.S. government information.

Authorization to photocopy items for internal or personal use is granted by ILSI for libraries and other users registered with the Copyright Clearance Center (CCC) Transactional Reporting Services, provided that $0.50 per page per copy is paid directly to CCC, 222 Rosewood Drive, Danvers, MA 01923. Telephone: (+1) 508 750 8400.

The use of trade names and commercial sources in this document is for purposes of identification only, and does not imply endorsement by the International Life Sciences Institute (ILSI). In addition, the views expressed herein are those of the individual authors of the original book and/or their organizations, and do not necessarily reflect those of ILSI (or the author of the concise monograph).

ILSI Press
1126 Sixteenth Street, N.W.
Washington, DC 20036-4810
USA
Telephone: (+1) 202 659 0074
Telefax: (+1) 202 659 8654

ILSI Europe
Avenue E. Mounier 83, Box 6
B-1200 Brussels
Belgium
Telephone: (+32) 2 771 00 14
Telefax: (+32) 2 762 00 44

Printed in Belgium

ISBN 0-944398-86-3

# FOREWORD

The use of microwave energy to heat and to cook foods has revolutionised culinary habits in recent decades. The continued success of microwave ovens is to be seen in their rapid expansion and wide distribution in both family and collective kitchens as well as in the increasing number of microwavable foods available.

Answers to bona-fide questions on the "whats" and the "hows" of the microwave mechanism, as well as on the risks involved in ingesting microwave-treated foods, have not always reached the consumer at large. Thus, ignorant of the current status of scientific knowledge of microwaves and their applications, the public still questions their innocuity.

In this context, the ILSI Europe Microwave Oven Task Force has contributed to demonstrating the safety of microwave-treated foods and the preservation of their nutritional quality. It was thought that the major responsibility of the task force was to provide a comprehensive guide with background material on all relevant aspects of microwaving foods: heating process, oven description, food and packaging characteristics, heat distribution, microbiology, nutrition and safety. This concise monograph, which integrates all of these aspects, certainly meets a need, and can be read by persons with various levels of training in the basic food and life sciences.

*Text developed by:* Allison Hill
*in collaboration with:* The ILSI Europe Microwave Oven Task Force
*Scientific Editor:* Philip Richardson
*Scientific Referees:* Klaus Werner Bögl and Andrew C. Metaxas
*Concise Monograph Series Editor:* Nicholas J. Jardine

# CONTENTS

# INTRODUCTION

Today microwave ovens are an established feature in millions of kitchens throughout the world. Yet the development of the microwave oven is relatively recent. During World War II the pioneers of microwave communication systems discovered that microwaves have many uses. Scientists at early warning stations found that when birds collided with radar masts they dropped to the ground, sizzling and well cooked. From this emerged the idea that food could be heated by microwaves.

The possibilities raised by this accidental discovery were examined further and before long microwave ovens could be found in both hospital and army mess kitchens. Since then, microwave ovens have become increasingly sophisticated, as have the products and packaging materials developed for microwave use.

This concise monograph summarises the current status of scientific knowledge about microwave ovens and their uses. The first part of this concise monograph describes how a microwave oven works and explains how products and packaging have developed to optimise the benefits of this particular method of heating foods. The second part examines safety and nutrition, the two issues which have been the focus of the main concerns raised about the application of this technology in food preparation.

# HOW MICROWAVE OVENS WORK

## The basic structure of a microwave oven

Figure 1 illustrates a typical microwave oven. It consists of a cavity surrounded by metal walls with a door at the front. The microwaves, generated by a magnetron, enter the cavity via a metallic tube called a wave guide. The stirrer helps to distribute the microwaves while the turntable improves energy distribution in the food.

### The heating process

Microwaves form part of the electromagnetic spectrum as do television and radio waves, infrared radiation, ultraviolet radiation and visible light. These electromagnetic waves are a means of transmitting energy through space just as electricity is electrical energy transmitted through a wire. Two frequencies have been set aside for use in microwave heating applications. These are 915 MHz (896 MHz in the UK) and 2450 MHz. Domestic appliances operate on 2450 MHz, whilst industrial applications make use of both frequencies (Figure 2).

The magnetron in the oven converts electrical energy at low frequencies into an electromagnetic field that oscillates at a much higher frequency.

Food is able to absorb microwave energy, later releasing the stored energy as heat. The main mechanism of heating occurs as a result of rotation of water molecules brought about by the microwaves. Water is a molecule which is unevenly charged along its structure, being more negatively charged in some parts of the molecule

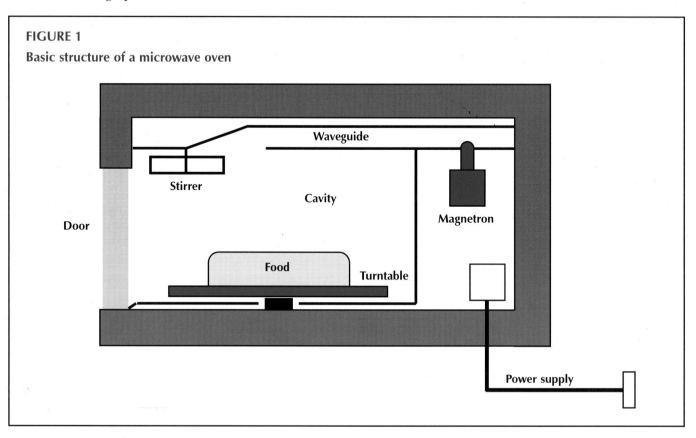

**FIGURE 1**

**Basic structure of a microwave oven**

and more positively charged in others. Such molecules are called "dipoles". When exposed to an electromagnetic field, dipoles behave as if they are tiny magnets and try to line up with it.

The electromagnetic field generated in a microwave oven oscillates with a frequency of 2450 MHz (or 2450 million cycles per second), so these dipoles (or molecular magnets) have to move back and forth at a speed of over two thousand million times a second to follow the field (Figure 3). Energy is absorbed in this process affecting the molecules themselves and their

interactions with neighbouring molecules. The energy of the microwaves is released in the form of heat, the amount produced being proportional to the energy absorbed by the food.

Other dipole molecules are involved in the heating process, but for food the most important dipole is water. Although some foods have a water content of zero (for example, the sugar, sucrose), others have up to 96%. However, there is also a contribution to heating from ionic material in food. Ions are also accelerated by the microwave's electromagnetic field (Figure 3) and collide

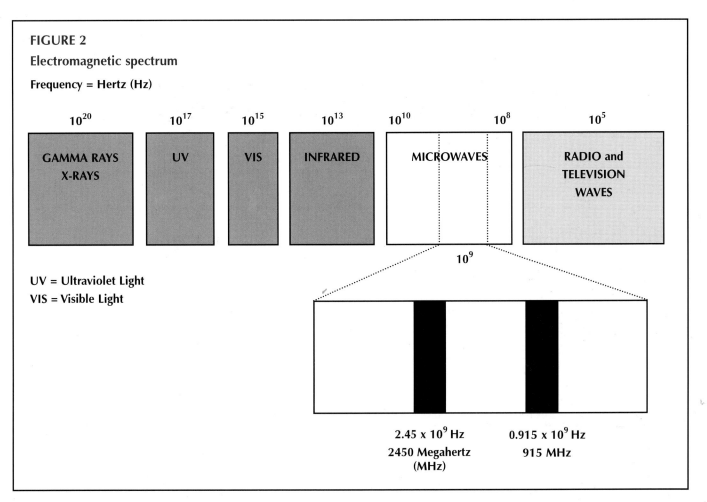

**FIGURE 2**

**Electromagnetic spectrum**

**Frequency = Hertz (Hz)**

| $10^{20}$ | $10^{17}$ | $10^{15}$ | $10^{13}$ | $10^{10}$ | $10^8$ | $10^5$ |

GAMMA RAYS
X-RAYS | UV | VIS | INFRARED | MICROWAVES | RADIO and TELEVISION WAVES

UV = Ultraviolet Light
VIS = Visible Light

$10^9$

$2.45 \times 10^9$ Hz
2450 Megahertz
(MHz)

$0.915 \times 10^9$ Hz
915 MHz

with other molecules to contribute to the heating process through the energy of the collision. This is important for foods with a high salt content. Molecular movement is also possible in fats, so these too can heat up in the presence of microwaves.

*Paper*

## Penetration of microwaves into food

The depth to which microwaves penetrate differs for each type of food. As the degree of absorbency in a food increases, the penetration depth decreases. Therefore, the greater the liquid water or salt content, the more limited is the depth where the influence of microwave heating is felt, and the greater the heating near the surface. Subsequent heat conduction and convection are

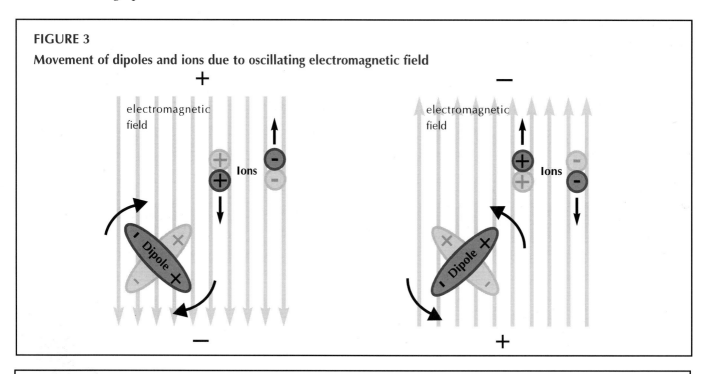

**FIGURE 3**

**Movement of dipoles and ions due to oscillating electromagnetic field**

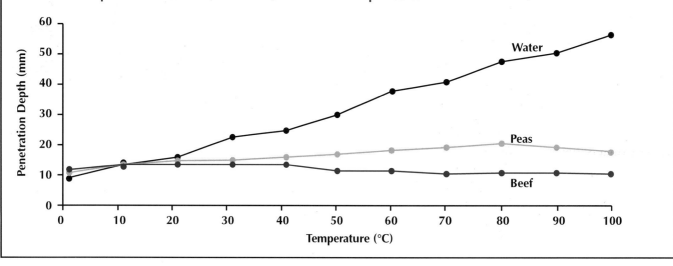

**FIGURE 4**

**Penetration depth of microwaves (2450 MHz) at different temperatures**

responsible for the distribution of the heat throughout the food product.

Penetration depth is also dependent on frequency and temperature. It is greater at 915 MHz than at 2450 MHz, which is why the lower frequency is the favoured one for many industrial applications such as thawing large blocks of frozen material. Both frequencies have been used for other applications such as drying and bulk cooking.

Data on penetration depth has been extensively tabulated for many food components. This information helps in the prediction of microwave heating rates. Figure 4 illustrates the variation of penetration depth with temperature.

## Temperatures generated in microwaved food

The electromagnetic field distribution generated in the oven cavity is not uniform and is affected by the product placed in it. This generally leads to uneven energy distribution and can cause hot and cold spots in the product. These hot and cold spots can be minimized by using stirrers which distribute the microwave energy more uniformly, or by rotating the food in the microwave oven on a turntable (Figure 1).

Microwave absorption is only one indicator of how hot a material will get in a microwave oven; materials such as fats will heat very rapidly even though the level of microwave absorption is relatively low in comparison with water. This is because fat has only a low heat capacity (that is, the amount of heat required to raise the temperature by a given amount). Water on the other hand has a relatively high heat capacity and heats more slowly than might be expected considering the relatively high microwave absorption it shows.

For foods of high water content which are cooked or reheated in microwave ovens, the temperature at the surface is generally no greater than 100°C. If the heating is extended until the material loses all of its water then higher temperatures can be achieved. For foods of high

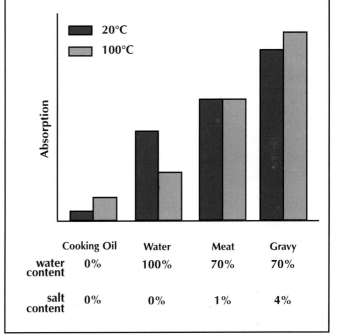

**FIGURE 5**

**Relative absorption of microwaves by different food products and components**

Note how temperature affects absorbancy of microwaves, as does the presence of ions (from salt)

fat content temperatures in excess of 200°C can be observed. This is seen when cooking bacon. Foods with a high sugar content, such as toffee-like foods, can reach 150°C.

Figure 5 illustrates the variability of microwave absorption for different food products. The temperature dependency of the absorption of microwaves in water and salt can also be seen.

## Heating frozen foods

The absorption characteristics of a food or ingredient depend on its physical state. An example of this is the much higher microwave absorption exhibited by water compared to ice. This is important when heating frozen foods. As parts of the product melt (due to the uneven nature of microwave absorption), heating greatly increases in that part of the food. In some cases the temperature in this region will increase rapidly whilst other parts of the food remain frozen leading to "thermal runaway" in thawed parts of the product.

Figure 6 shows the relative difference in microwave absorption below and above 0°C for a variety of food materials.

In industrial processing to avoid the problem of thermal runaway, frozen products are "tempered". They are brought to a temperature of -3°C at which point they can be manipulated rather than thawed through 0°C. Thawing without thermal runaway can also be achieved by pulsing the microwaves on and off.

## The oven

### Oven structure

As shown in Figure 1, a typical microwave oven consists of a cavity surrounded by metal walls with a door at the front. The metal walls reflect the microwaves. To facilitate uniform heating of the food by the microwaves, it is necessary to establish as many "modes" as possible inside the oven. A mode is a specific pattern of electromagnetic waves set up in the oven. The dimensions of the inside cavity are chosen so as to maximise the number of modes. The cavity size is not critical as evidenced by the different sized ovens available; each size supports different numbers of

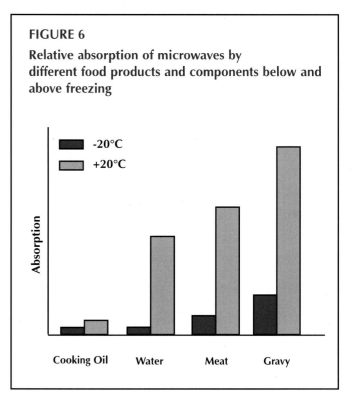

**FIGURE 6**

**Relative absorption of microwaves by different food products and components below and above freezing**

modes which typically range from 30 to 50. In some models a mode stirrer also helps to distribute the electromagnetic energy, while in others a turntable is used to improve energy distribution in the food; many models have both.

The multiplicity of modes and their interaction in such "multimode" ovens ensures that the energy transferred from the source to the cavity is absorbed more efficiently by the food. The origin of hot and cold spots in the food is the manifestation of the combined action of the peaks and troughs of the many modes inside the cavity.

The door is designed to ensure that the cavity is totally sealed so that no microwaves can escape; this is achieved by using mechanical locks and a wave trap, commonly known as a "choke" seal, which is usually fitted with microwave absorbing material.

The ultra-short wave generator, the magnetron, is enclosed in a housing to which only technical personnel have access. The microwaves are emitted either directly into the cavity or via a tube called a wave guide.

The front of the oven bears an "on" switch, a time switch and a switch for selecting the required power. More expensive models carry a greater range of facilities for carrying out different cooking routines.

## Physical safety

Microwave ovens are designed to ensure that microwave energy is confined within the cavity and does not leak outside the oven. This is important because if microwave energy does escape, it can be absorbed by the human body, turned into heat and can damage the body's soft tissues. Strict international standards control the safety of microwave ovens, and most manufacturers have even stricter standards. Furthermore, consumer associations periodically check ovens for leaks and publish the results. Given these stringent checks, the only realistic danger is accidental damage to the grid, which is protected by the glass door. This should be checked at regular intervals.

## Performance testing

The International Electrotechnical Commission sets out standards (IEC 705) for measuring the performance of microwave cooking appliances. They are designed to ensure consistent performance standards irrespective of the manufacturer or country of manufacture.

IEC 705 includes a detailed procedure for determining the power output of ovens. In simple terms, this is achieved by taking a glass container of water at a known temperature and heating it to a higher temperature and measuring the time taken. Various tests using water in cups and in tanks divided into individual compartments are also used to test the heating uniformity of ovens. Further detailed tests also provide a means of establishing the cooking and defrosting performance of ovens for a range of foods.

## Food containers for microwave ovens

There are two categories of food containers: those used for prepackaged foods and those used for oven-proof cookware. In the case of most prepackaged foods the container is expected to have a single use. In contrast, oven-proof cookware containers are purchased empty of food and are used in the home for cooking or reheating foods; they are expected to have repeated use.

The choice of materials for microwave packaging is strongly influenced by the desired effects of the microwave-container interaction.

## The use of microwave-transparent materials

These include china, glass, paper and plastics, which are all transparent to microwaves and so do not heat directly in the microwave field. Most packaging materials are of this type because they allow maximum microwave energy to be absorbed by the food.

Glass retains its heat, so care must be taken when handling microwave-heated foods in glass containers. Also, some kinds of glass are more absorptive than others. For jars with a small radius, microwave energy can be superimposed at the centre, giving rise to scalding temperatures in that area while the outer regions are only moderately warm. Stirring the food partway through cooking can help overcome this.

Plastic cartons can prevent contamination through the distribution chain, are lightweight and are relatively inexpensive. They can be produced in a variety of shapes and colours, and most can be used straight from the freezer to the microwave oven and then to the table.

High-density polyethylene is generally used for high-water-content foods, but cannot be used for high-fat or high sugar content foodstuffs where temperatures exceeding 100°C may be reached. The most commonly used packaging materials are polypropylene and crystalline polyethylene terephthalate (CPET), which have melting points of 210–230°C and are therefore suitable for most foods. CPET has the added advantage of being suitable for both microwave and conventional oven use.

Consumers are advised not to use plastic containers such as margarine tubs or yogurt containers in microwave ovens because these materials are not manufactured for high-temperature use.

Plastic films are also produced for use in microwave ovens. When hot, substances used in the manufacture of the plastic film may migrate from the film into the food. Consumers are therefore advised to follow manufacturers' instructions on the use of such films. Plastic films should not be used in conventional ovens for any purposes.

Paper and board are inexpensive, well-accepted materials, but they have strength limitations, particularly when wet. They also absorb some of the microwave energy.

## Metals in microwave ovens

The use of metals in microwave ovens requires some care. Many will have observed the sparks caused by electrical arcing after inadvertently putting crockery with a gilded rim into a microwave oven. Such arcing should be avoided as it can damage the magnetron. However, smooth metallic surfaces reflect microwaves and do not heat; that is why the oven cavity has metal walls, usually stainless steel. A metallic material can, therefore, be used to shield quick-heating foods so that a more uniform heating of different product components can be achieved. Likewise, foil containers can be used in the microwave oven provided they are either unlidded or have microwave-transparent lids. Since the energy is only absorbed from the top, problems of overheating at the edges can thereby be avoided, although more time will be needed for effective heating to take place. See Box 1 for more detailed advice.

## Browning and crisping in microwave ovens

Most microwave cookware is microwave-transparent and as such does not get hot. The exceptions are microwave browning dishes, and the group of microwave packaging materials called microwave

## BOX 1

### Guidelines for using metal containers in microwave ovens

- Use a shallow container with a large product surface area and ensure that it is two-thirds full.

- Remove the lid (unless it is microwave-transparent) and place the container on a non-metallic, heat-resistant tray or turntable in the centre of the oven. If the oven turntable is metal, place the container on an upturned oven-proof plate or microwave insulating tray.

- Use foil containers singly in the oven and do not use damaged or torn metallic containers.

- Do not allow a metal container to touch the cavity sides. Ensure foil lids are completely removed.

- Do not use a microwave oven containing sharp metallic implements like knives, forks, spoons etc. Do not use china with metallic decoration.

- Allow greater time for the food to heat compared to foods in non-metal containers, and ensure that it is piping hot before serving.

susceptors. These components are designed to absorb microwaves and get hot quickly to produce crisping effects. They reach temperatures of 250°C in normal use.

A typical microwave-active material used for this purpose is aluminium which is sputtered as minute particles onto a thin layer of polyester film. Compared to smooth metal, the particulate nature of the aluminium in this case permits absorption of microwave energy, causing it to heat to the required temperature. The aluminised polyester sheet is glued aluminium-side down onto thin paper board. The polyester layer prevents contact between the aluminium and the food (Figure 7). Susceptor materials, sold on a roll, are available for browning and crisping pastry products. Ceramics incorporating microwave-absorbing materials can also be used as susceptors.

## Cooking foods in microwave ovens

It is clear that the heating process itself and the microwave interaction with the food container both affect how different food components are heated. This in turn influences product design. The following factors need to be taken into account to gain the best results when preparing food for the microwave oven.

### Composition

The level and distribution of water, oils and salts affects the way a product heats. The addition of salt for example, at the surface of the food causes the surface to heat well but decreases the penetration depth.

### Size

Small items cook faster than large items. When a large sized foodstuff like a turkey is heated, most of the microwave energy is dissipated before it reaches the centre. The central layers are heated by thermal conduction but this is a much slower process than microwave heating. Prolonged exposure of the food to the microwave field to achieve the desired thermal effect can result in a dried and tough outer layer. To compensate, a "standing time" allows the thermal energy to be dissipated throughout the food after the microwave energy has been turned off. This is effective

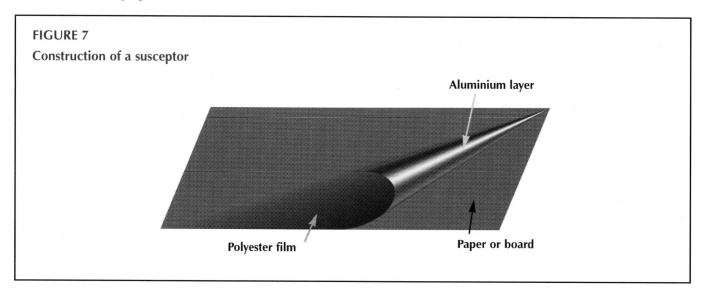

**FIGURE 7**

**Construction of a susceptor**

Aluminium layer

Polyester film

Paper or board

with meat products, which conduct heat well, but is of little effect for denser products. In the latter case, restricting product size or using combination ovens are the preferred solutions (see below).

## Quantity

Because it takes longer to heat multiple items than single items, more time should be allowed to heat more than one product or package at a time in the microwave oven.

## Shape

The shape of each food component must be considered, whether cooking a single item or several items together. Irregularly shaped products like chicken legs, wings, and thighs heat nonuniformly; the thinner, narrower parts tend to heat more rapidly and may overcook or dry out by the time the thicker parts are sufficiently heated.

A spherical shape will, subject to the penetration depth, tend to be hotter toward the centre of the product. To allow for this "focusing effect", extra standing time or stirring the food may be necessary to allow the temperature to distribute evenly throughout the product as a whole. In addition, foods sealed by a membrane (for example, sausages, eggs and certain vegetables) should be pierced to prevent them from exploding.

A large surface area in relation to volume will result in a rapid heating rate because of the greater area available to interact with the microwave field. Thus, changing the surface area can control the rate of heating. For example, chopping up certain foods can increase the rate of heating.

## Density

Foods which are dense generally take longer to heat than those which are more open and porous. A thick piece of steak may be the same size and shape as a hamburger, but the former consists of varying formations of lean meat, fats and gristle. The steak can become shrivelled and cooked on the surface while the interior is far from done; the hamburger, by contrast, has its fat uniformly distributed throughout, resulting in more even heating.

### Surface effects

The temperature of the microwave oven atmosphere is not affected by microwaves. Because the food heats rapidly, water vapour is driven out of the core and condenses at the surface. In the case of bread and pizza products this effect is characterised by soggy and uncooked crusts with poor organoleptic properties. In the case of meat products the result can be dried and tough outer layers. These problems can be overcome by using susceptors to aid the heating process or, in some cases, by reducing the product size.

### Component layout in food products

Product layout has far greater importance for microwave than for conventional heating methods and can be used as a successful measure of ensuring that a multicomponent product reaches a uniformly high temperature. The thermal, electrical and dielectric properties of each component must be considered by food product manufacturers. Care should be taken in the layout of the components of a prepared meal to avoid shielding effects (and thus inadequate heating) where these are not desired. Shielding can be incorporated into the packaging of microwavable foods to ensure that sensitive or rapidly heated areas are protected.

## *Comparison with conventional ovens*

### Heat distribution

The key difference between microwave heating and cooking in a conventional oven is the way in which the heat is generated and distributed. In a microwave oven food is heated directly by the microwaves, whereas in a conventional oven food is cooked by the heat from the surrounding environment. Food heated in a microwave oven is subjected to a heterogeneous temperature pattern with the potential for hot and cold spots. In a conventional oven the surface temperature of the food is very high (close to that of the oven atmosphere, which might range from 150 to 250°C) compared with that of the interior, which generally peaks at about 90–95°C. In microwave-heated foods the inner food core may be either at a higher or at a lower temperature than the outer parts, depending on the size and shape of the product and on its water and salt contents (see Figure 8).

Improvements in microwave oven design have minimised the problems of heterogeneous heating. Allowing the heated product to stand for a short time before consumption also helps. Other improvements include humidity and heat sensors which are effective in controlling the heating process to avoid both under and overcooking/overheating.

The potential for low temperatures and non-uniformity of temperature distribution has given rise to concerns relating to nutrition and food safety. These are dealt with in the Safety and Nutrition section.

**FIGURE 8**

A comparison of the temperatures of the oven, food interior and food surface for a conventional oven and a microwave oven with and without susceptors for a typical food product

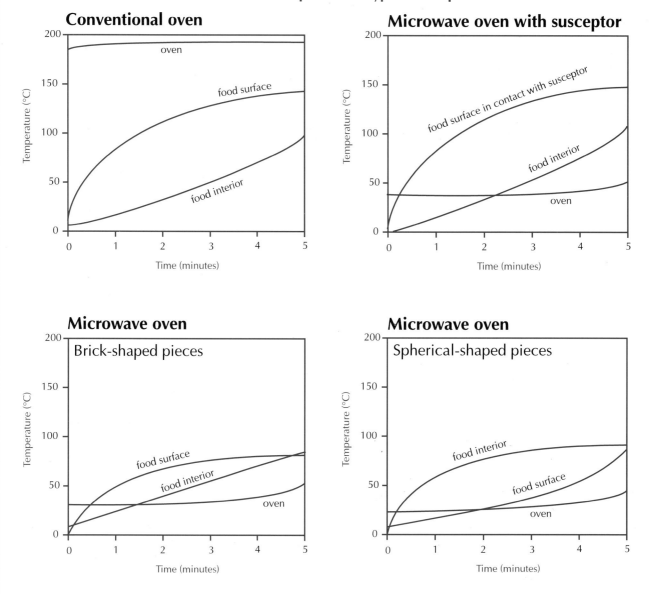

### Flavour/texture

The biggest difference between conventional ovens and microwave ovens is the inability of microwave ovens to induce browning or crisping of foods, although this can be achieved in microwave-heated foods by using susceptors. Alternatively, combination ovens use microwaves to achieve volumetric heating and conventional heat sources for surface browning or crisping.

There is a range of opinion about the flavour of microwave versus conventionally cooked foods. Opinions often depend on the type of food. For example, microwave cooked fish and vegetables are usually considered to have a better flavour. In contrast, many believe casseroles have a better flavour when cooked conventionally because the longer cooking time allows the flavours to develop more fully.

### Efficiency

*Energy savings.* As conventional ovens heat the air surrounding a food, a certain amount of thermal energy escapes to the outside air; this is also true of combination ovens. In the case of microwave-only ovens, heat is generated within the food and little thermal energy escapes from the oven.

Compared with conventional electric or gas appliances, home microwave ovens showed average energy savings of 21% for nine bakery products, 29% for five broiled meats and 22% for rib roasts. In contrast, turkey required slightly more energy. Generally, microwave ovens consume as much or more energy than electric cooktop ranges in the preparation of a single meal item but less than a conventional electrical oven.

*Time savings.* Microwave ovens offer clear time savings over conventional ovens. Depending on the product, most microwave ovens require only about 20% of the time used by conventional heating systems. However, when more than one portion of food is to be cooked or reheated, the time savings are somewhat reduced. Two portions heated in a microwave oven will require nearly twice as much time as one portion, whereas in a conventional oven, multiple amounts require little extra time. Nevertheless, the time taken overall will still be less with microwaving than with conventional cooking.

## SAFETY AND NUTRITION

### Microbiology

All fresh foods are contaminated to some degree by microorganisms; some are harmless, whereas others are dangerous. The growth of microorganisms in foods is stopped by deep-freezing and slowed down by refrigeration, and microorganisms can be killed by strict heat treatments such as pasteurization and sterilization. Traditional cooking processes practised in the home, that combine high temperatures with relatively long cooking times, also help kill microorganisms.

The question has arisen whether microwaving is effective in killing microorganisms. Concerns are based on the shorter times employed in microwave cooking and in the known heterogeneous nature of the heating process. Compared with conventional cooking, microwave heating might result in lower temperatures at the food surface; because this is where most foodborne microorganisms are found, this is a logical concern.

Many studies have addressed this question. The scientific consensus is that microwave ovens are as effective as conventional ovens in killing microorganisms, providing the appropriate temperatures are reached. For example, to kill *Listeria monocytogenes* it is commonly advised that the coolest part of the food should be held at 70°C for not less than 2 minutes, or an equivalent time/temperature combination. Studies have shown that this will reduce the number of viable microorganisms by a factor of at least $10^6$. The survival of microorganisms in microwaved meats above this temperature has been attributed to uneven heating resulting in cold spots.

Similarly, studies have shown that *Trichinella spiralis*, a parasite which causes trichinosis in humans, survived in microwave cooked pork to centre temperatures of 77°C, although viable trichinae were not found in pork cooked in conventional ovens at centre temperatures of 58°C. However, when pork inoculated with massive larval doses of the organism was microwaved in plastic roasting bags producing a much higher surface temperature, the trichinae did not survive.

The heating process therefore needs to take into account the different heating effects of microwave and conventional ovens, to ensure that adequate temperatures are reached. A number of testing procedures have been developed to ensure that microwave ovens will heat foods to stipulated temperatures.

Thermal image analysis and fibre optic probes are used to measure temperatures at the food surface and within the food, respectively. Another test, known as the alginate bead technique, enables the actual effect of lethal temperatures on the microorganism to be determined. This is done by immobilising microorganisms with known heat sensitivity in alginate beads, which are then placed throughout the food prior to microwaving. The lethality of the times and temperatures achieved during microwave cooking at multiple points throughout a product can then be measured accurately. Based on these tests, heating instructions can be provided to the consumer to enable a food to be heated sufficiently to kill microorganisms.

Above all, it remains important that all products, whether destined for the microwave or the conventional oven, should be processed, packaged, distributed and stored in ways that minimise the presence and growth of microorganisms.

Concerns have been raised about the athermal effects – effects different from heating effects – of microwave ovens. These include changes in microorganisms at sub-lethal temperatures. However, all cases where such an effect has been identified have raised questions about the effective measurement and control of temperature used in the cooking process. The current consensus is that the interaction of microwaves with microbiological systems does not produce athermal effects.

## *Carcinogens and mutagens*

It is known that cooking can lead to the production of carcinogens and mutagens in certain foods. The question arises of the extent to which microwave cooking increases or decreases these reactions.

It has been found that mutagen production is a function of the product as well as the cooking time and method (Box 2). Cooking meat (especially at high temperatures) can result in the formation of mutagenic/carcinogenic substances such as heterocyclic amines (carbolines, quinolines, quinoxalines), volatile nitrosamines and polycyclic aromatic hydrocarbons.

Studies have shown that mutagenic activity in grilled lamb chops, sirloin steaks and rump steaks was directly related to the average surface temperatures reached during cooking. The use of butter for frying was

# BOX 2

## Microwaves and the formation of mutagens

### Comparison of microwave and conventional cooking

#### Heterocyclic amines (HAs)

Precursors of HAs are found throughout meat. Thus, searing the exterior of a steak will produce few, if any, HAs. But as heat begins to drive water and fat out, key ingredients in HA formation move toward the surface. Studies have shown that precooking burgers for 2 minutes in a microwave oven drives off some juice – as well as most of the precursors of HAs. When later barbecued, these burgers produce only about 10% of the carcinogens seen in burgers cooked solely on a grill. The type of protein is also significant. A study comparing ground-meat burgers fried identically, indicated that chicken and beef produce five to eight times more HAs than does fish.

#### Nitrosamines

Bacon was analysed for volatile nitrosamines, which are known carcinogens, after microwave cooking, and the results were compared with those obtained after frying bacon in a pan. Microwave cooking gave statistically significantly lower levels of all three volatile nitrosamines detected in the bacon; for example, N-nitroso-pyrrolidine, NPYR, was found in all 20 samples fried in a pan but in only five microwave-cooked samples. The authors concluded that the lower level of nitrosamines resulting from microwave cooking was probably the result of the lower cooking temperature and the short exposure time to the heat.

#### Polycyclic aromatic hydrocarbons (PAHs)

A study of beef cooked in corn oil analysed for PAHs found that conventional frying and reheating generated substantial levels of PAHs whereas microwave cooking and reheating did not generate significant levels of PAHs in the meat.

particularly associated with higher mutagenicity in meat samples. Fried meats generally produced more mutagens than did grilled meats at comparable temperatures.

In contrast, microwaving, along with boiling and stewing, resulted in little or no mutagenicity, probably because in each case the temperatures did not rise above 100 °C. In one study of beef, microwave cooking caused no mutagenic activity even when the food had been heated three times longer than the normal cooking period, to the point of inedibility resulting from browning and hardening.

Time also affects the quantity of mutagens formed during pan broiling, although temperature remains the more important factor. Lowering the cooking temperature as much as possible is advised to minimize the formation of mutagens when frying or broiling meat. An effective way of shortening the cooking time is to brown meat lightly in the frying pan and to complete the cooking in either a microwave or a regular oven.

## BOX 3

### Nutritive values of proteins in microwaved foods

Numerous studies using different methodologies, have considered how heating protein foods by microwaves affects the nutritional value of the proteins. No detrimental effects have been shown. A study compares the effects of heating beef pot roast and beans with frankfurters on the availability of lysine and methionine, the two most important essential amino acids in food proteins. The study also looked at instant mashed potatoes, peas with onions, and breaded fish. For all foods it was found that neither conventional nor microwave heating seriously decreased the nutritive value of the protein as measured by chemical analysis and by human and animal trials.

In another study, the nutritive value of a casein solution heated by microwave or conventional methods was determined in a feeding study of rats. The net protein utilisation, digestibility and biological values of unheated casein were 61.1%, 100%, and 61%, respectively. Comparable values were obtained for casein heated by microwave or conventionally. The authors concluded that there are no negative effects on protein nutritive value by either microwave or conventional heating under conditions likely to be encountered in household situations.

It has also been shown in well-designed studies carried out by a variety of laboratories that there are no adverse effects on milk or infant formulas when heated in microwave ovens. This is true whether normal domestic conditions were used, or higher but controlled temperatures and longer duration.

## Nutrients

### Proteins

When proteins are heated they are denatured; that is, their molecular structure is modified. This limits their physical properties such as solubility and their biological properties including their enzymic and immunologic activities.

There are slight differences in denaturation rates in foods heated in the microwave compared with conventional heating, because of differences in heating time and temperature. In general, however, the nutritive value of proteins in foods is comparable, whether the cooking is done by microwaves or conventional means.

Studies on thawing of human plasma and breast milk show that when the microwave treatment is well controlled to avoid overheating, such biological properties as the clotting activity of plasma and the antibacterial activities of human milk are reasonably maintained. For further information, refer to Box 3.

### Lipids

The effect of microwaves on lipid stability has also been investigated. The hydrolysis of triglycerides has been studied in soya, egg yolk and meats. Fatty acid stability has been studied in meat, chicken fat, beef tallow, bacon fat, rainbow trout and peanut oil. The peroxidation of polyunsaturated fatty acids, which generates rancidity, has been studied in meat, egg yolk and chicken. The conclusion is that microwave cooking does not induce more chemical modifications in fats than do conventional cooking methods. In particular, microwave cooking does not generate free radicals because, compared with conventional treatments, it does not result in more lipid oxidation.

## Vitamins

Numerous studies have been carried out to determine the stability of vitamins in microwaved foods, including a wide range of meats and vegetables, compared with their stability in conventionally cooked foods.

Retention of fat-soluble and water-soluble vitamins varies with cooking time, internal temperature, product type, and the size, type and power of the oven. In general, vitamin retention in microwaved foods is the same as in foods cooked conventionally; in some cases it is even better. Foods cooked or blanched at higher energy levels are heated more quickly and, if heated for a shorter length of time, retain greater amounts of heat-labile nutrients than do foods heated by other methods. The Institute of Food Technologists' Expert Panel on Food Safety and Nutrition concluded that vitamin retention in microwaved foods is improved because cooking time is shortened.

## Minerals

Minerals cannot be destroyed during food treatment; they can, however, be lost in cooking water or meat drippings. Studies of a variety of vegetables cooked conventionally, and in a microwave oven, for various

## BOX 4

### A toxicity study of microwaved food

A major study was carried out to examine the potential toxicity of microwaved food. This consisted of a standardised subchronic (i.e., short-term, in this case 3 months) toxicity trial in rats fed diets composed of a range of microwave-cooked and, for comparison, conventionally cooked human foodstuffs representative of the normal European diet (beef, potatoes, and a mixture of cereals, legumes and root vegetables, plus vegetable oil).

The cooked ingredients were freeze-dried, ground and mixed with supplements of vitamins and minerals to meet the rats' nutritional requirements. An additional control group was fed a cereal-based rodent diet. Criteria to assess toxicity included clinical observations, ophthalmoscopy, growth, food and water intake, haematology, clinical chemistry, urinalysis, organ weights, micronucleated erythrocytes in bone marrow, gross examination at autopsy and microscopic examination of a wide range of organs. The rats fed on microwaved foods suffered no adverse effects compared with those that ate conventionally cooked foods.

The food was also subjected to misuse treatment by repeated reheating. This was designed to increase the concentration of any potentially harmful substances which could be present in the food, albeit previously undetected. Again no harmful effects were found.

lengths of time, revealed a negligible change in mineral content regardless of cooking method.

In one study of roasts of beef, phosphorus and potassium were retained significantly more when cooked by microwave than when braised in a conventional oven. Another study found significantly higher retention of some minerals after conventional cooking.

In the case of both vitamins and minerals, where nutrients do leach from the food, they can be recovered when the water or dripping is used, for example, for gravies.

## Packaging migration

It is known that under certain conditions very small amounts of material can migrate from plastic packaging into food when heated. Strict legislative controls therefore specify substances which can be used in the manufacture of food contact plastics, and limits are placed on the total amount of material which can migrate from a plastic.

Plastic films are widely recommended for use in microwave cookery to cover foods during heating to prevent the surface from drying out. They are often proposed for baking, reheating plates of precooked food, wrapping whole large vegetables during cooking and for covering frozen dishes during reheating directly from the freezer. A number of recipes call for foods to be cooked in direct contact with film. These foods range from meat, poultry, fish and vegetable dishes to steamed puddings, cakes and biscuits.

Testing of polyvinyl chloride (PVC) film ("cling film") plasticized with di (2-ethylhexyl) adipate (DEHA) was carried out in the United Kingdom in 1986 for a variety of foods which were either cooked or reheated in microwave ovens. DEHA migration from these films was found in a variety of situations. This led to the recommendation that such film should not be used in any circumstances in conventional ovens, nor should it be used for lining dishes or wrapping foods in a microwave oven. Reheating of foods covered with film in a microwave oven was, however, found to be acceptable, because migration was significantly lower.

## TABLE 1

### Migration of ATBC into microwave-cooked foods

| Food | mg/kg |
|------|-------|
| Soup, pork chop, sweet corn, brussels sprouts, steam pudding | 0 – 3 |
| Chicken breasts, hot bread, spinach, cakes, chocolate cake | 10 –25 |
| Peanuts, biscuits | 80 |

Many other studies have demonstrated that for a number of different plastics and for different compounds, no quantitative effect on migration could be observed as a result of microwaving foods over and above that expected from the heating effect alone.

In a study of the migration of the plasticiser acetyltributyl citrate (ATBC) into a variety of microwave-cooked meals, the lowest level of migration occurred where there was no direct contact between film and food (Table 1). The highest levels of migration were found where the film was used as a liner or wrapping in direct contact with a fatty food surface.

As long as the guidelines are followed, microwave use does not give rise to increased risks from packaging migration, although in some cases consumers can detect off-flavours when plastic materials come into contact with a heated food.

# CONCLUSION

The safety and nutritional issues raised by the introduction of microwave ovens have been extensively examined by internationally recognised experts in various institutes and research groups. International and regulatory bodies have concluded that heating food in a domestic microwave oven is not harmful.

Specifically, in February 1992 the World Health Organisation stated that it had "no information to support the contention that cooking with microwaves induces any toxic substances or harmful effects unique to this process .... There is no scientific evidence that foods prepared in microwave ovens present any health risk to consumers, provided that the instructions given by the manufacturers are followed."

# GLOSSARY OF TERMS

**Absorption:** The degree of take-up of microwave energy by the target (food) material

**Athermal effect:** Effects seen during microwaving which are not directly attributable to heating or thermal effects

**Biological Value:** A measure of the amount of absorbed nitrogen (protein) retained or used by the body

**Carcinogen:** A substance capable of initiating cancer

**Cavity:** The main part of the microwave oven in which food materials are placed for cooking or re-heating

**Digestibility:** A foods' capability for absorbtion after ingestion, sometimes given as a proportion or ratio

**Dipoles:** Charge imbalances within molecules which are susceptible to interaction with electrical fields

**Frequency:** The number of oscillations of an electrical field per second, usually quantified in hertz (Hz)

**Haematology:** Study of the blood and the blood-forming tissues

**Isomer:** Compound or ion that contains the same number of atoms of the same element, but differ in structural arrangement and properties

**Lethal temperature:** Temperature capable of causing death (of the microorganisms)

**Listeria monocytogenes:** A common species of bacteria that causes the infectious disease listeriosis

**Magnetron:** The electrical device within a microwave oven that generates microwaves

**Megahertz (Mhz):** A unit of frequency (1 Mhz = 1,000,000 Hz) used to measure wavelengths in the electromagnetic spectrum

**Micronucleated erythrocytes:** Red blood cells containing fragments of the original nucleus

**Microwave-active materials (susceptors):** Packaging materials or cookware which absorbs or reflects microwave energy, thus influencing the heating of food materials in the microwave oven

**Microwave-passive materials:** Packaging materials or cookware that permits the transmission of microwave energy to the target food material but does not absorb any microwave energy

**Microwave-reflective materials:** Materials that do not allow microwaves to pass through them, e.g., metals

**Mode:** A field configuration or pattern of guided waves, a state of an oscillating magnetron that corresponds to a particular field pattern

**Mutagen:** An environmental agent (chemical or physical) that induces a genetic mutation

**Net protein utilisation (NPU):** A percentage of the amount of ingested nitrogen (protein) retained or utilised by the body calculated by multiplying the digestibility of a food protein by its biological value

**Nutrient bioavailability:** The extent to which a nutrient is available to be used by the body

**Oven proof cookware:** Containers used for cooking or reheating food materials that can be placed inside both microwave and conventional ovens

**Penetration depth:** A measure of the depth within materials to which the effect of microwave heating is felt

**Power output:** The energy delivered by a microwave oven to a standard 1000 g water test load

**Racemisation:** The conversion of a population of a single optical isomer into a mixture of the two forms

**Standing time:** The time allowed after microwave heating has ended for the temperature to equilibrate within the food product

**Tempering:** The elevation of product temperature to just below the freezing point for further handling or processing

**Thermal conduction:** The movement of heat within a material by molecular interaction

**Thermal runaway:** The development of high-temperature pockets within frozen food materials when a part is thawed in the microwave oven but the remainder is still frozen

**Trichinosis:** Parasite worm infection, early symptoms of which include abdominal pain, nausea, fever and diarrhoea, followed by muscle pain and fatigue

**Turntable:** The device in the microwave oven that moves food material through the standing electrical field thus promoting even heating of the food product

**Waveguide:** The means of transmitting the microwave energy from the magnetron to the oven cavity itself

# ADDITIONAL READING

Bodwell, C.E, and Womack, M. (1978) *Journal of Food Science* 43: 1543–1549

Castle, L., et al. (1988) Migration of the plasticizer acetyltributyl citrate from plastic film into foods during microwave cooking and other domestic use. *Journal of Food Protection* 51 (December): 916–919

Cross, G.A., and Fung, D.Y.C. (1982) The effects of microwaves on nutrient value of foods. *Critical Reviews in Food Science and Nutrition* 16 (4): 355-419

Finot, P.A., and Merabet, M. (1993) Nutritional and safety aspects of microwaves: history and critical evaluation of reported studies. *International Journal of Food Sciences and Nutrition* 44 (Suppl., 1): S65–75

Jonker, D., and Penninks, A.H. (1992) Comparative study of the nutritive value of casein heated by microwave and conventionally. *J. Sci. Food Agric.* 59: 123–126

Jonker, J., and Til, H.P. (1995) Human diets cooked by microwave or conventionally: comparative subchronic (13-wk) toxicity study in rats. *Food and Chemical Toxicology* 33 (4): 245–256

Lorenz, K., and Decareau, R.V. (1976) Microwave heating of foods: changes in nutrient and chemical composition. *Critical Reviews in Food Science and Nutrition* 7 (4): 339-370

Marchelli, R., et al (1992) D-Amino acids in reconstituted infant formula: a comparison between conventional and microwave heating. *J. Sci. Food Agric.* 59: 217–226

Richardson, P.S. (1990) Microwave heating: the food technologist's toolkit. *Food Technology International Europe London*: Sterling Publications: 65-68